Too Cute!
Baby Penguins

WITHDRAWN

JOHNSTON PUBLIC LIBRARY
6700 MERLE HAY ROAD
JOHNSTON, IOWA 50131

by Betsy Rathburn

BELLWETHER MEDIA
MINNEAPOLIS, MN

Blastoff! Beginners are developed by literacy experts and educators to meet the needs of early readers. These engaging informational texts support young children as they begin reading about their world. Through simple language and high frequency words paired with crisp, colorful photos, Blastoff! Beginners launch young readers into the universe of independent reading.

Sight Words in This Book

and	find	new	they
are	from	now	this
at	have	one	to
can	in	the	water
down	like	their	
eat	look	these	

This edition first published in 2022 by Bellwether Media, Inc.

No part of this publication may be reproduced in whole or in part without written permission of the publisher. For information regarding permission, write to Bellwether Media, Inc., Attention: Permissions Department, 6012 Blue Circle Drive, Minnetonka, MN 55343.

Library of Congress Cataloging-in-Publication Data

Names: Rathburn, Betsy, author.
Title: Baby penguins / by Betsy Rathburn.
Description: Minneapolis, MN : Bellwether Media, 2022. | Series: Blastoff! beginners: Too cute! | Includes bibliographical references and index. | Audience: Ages 4-7 | Audience: Grades K-1
Identifiers: LCCN 2021001457 (print) | LCCN 2021001458 (ebook) | ISBN 9781644874882 (library binding) | ISBN 9781648344701 (paperback) | ISBN 9781648343964 (ebook)
Subjects: LCSH: Penguins--Infancy--Juvenile literature.
Classification: LCC QL696.S473 R375 2022 (print) | LCC QL696.S473 (ebook) | DDC 598.4713/92--dc23
LC record available at https://lccn.loc.gov/2021001457
LC ebook record available at https://lccn.loc.gov/2021001458

Text copyright © 2022 by Bellwether Media, Inc. BLASTOFF! BEGINNERS and associated logos are trademarks and/or registered trademarks of Bellwether Media, Inc.

Editor: Amy McDonald   Designer: Jeffrey Kollock

Printed in the United States of America, North Mankato, MN.

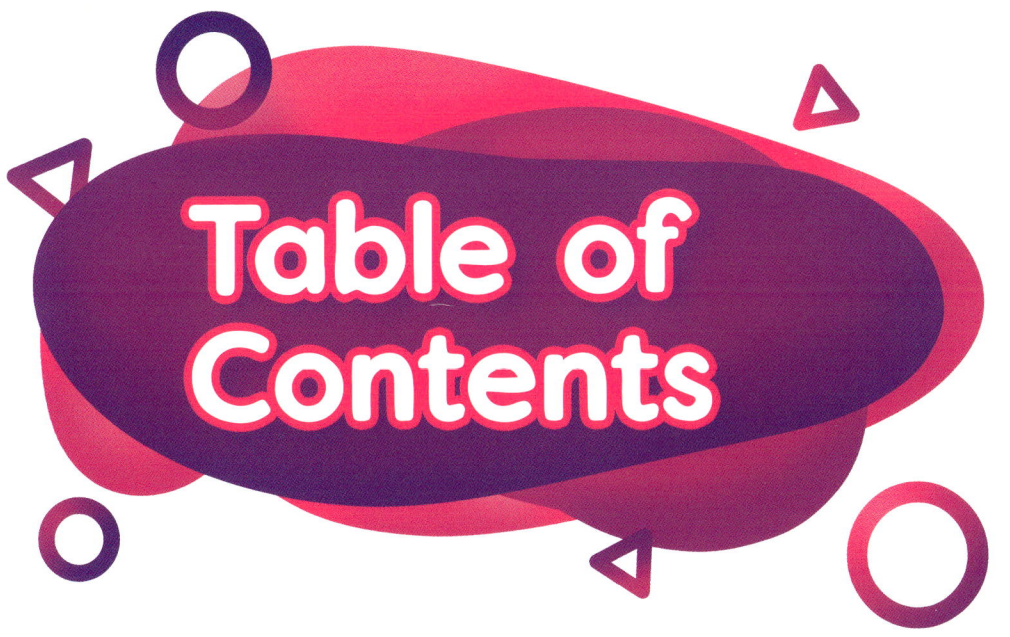

Table of Contents

A Baby Penguin!	4
Safe and Warm	6
All Grown Up!	18
Baby Penguin Facts	22
Glossary	23
To Learn More	24
Index	24

A Baby Penguin!

Look at the baby penguin. Hello, chick!

Safe and Warm

Chicks **hatch** from eggs. They live near water.

Chicks have fuzzy **down**. They keep warm.

down

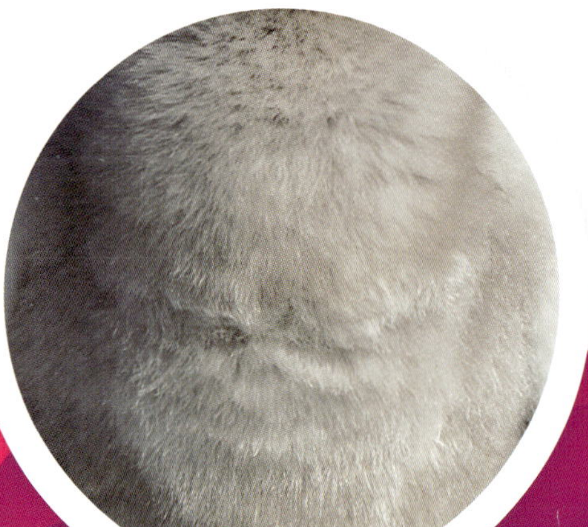

Chicks stay near mom and dad. They keep safe.

These chicks are hungry! Mom and dad find food.

Chicks like to eat fish. Mom feeds this one.

Chicks **huddle** together. They keep warm and safe. Stay close!

huddle

All Grown Up!

Chicks lose their down. They grow new **feathers**.

feathers

Now the chicks can swim. Dive in!

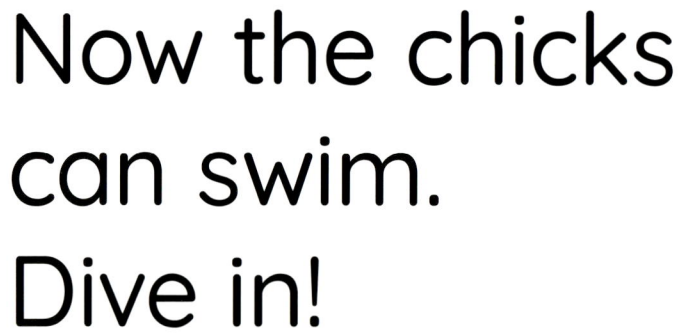

Baby Penguin Facts

Penguin Life Stages

egg   chick   fledgling   adult

A Day in the Life

eat   huddle   swim

Glossary

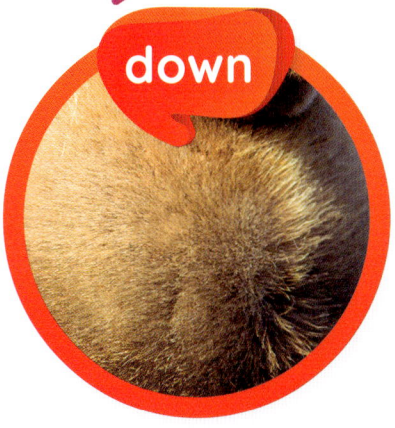

down

soft feathers that keep some birds warm

feathers

soft coverings that help penguins swim in cold water

hatch

to break out of an egg

huddle

to gather closely

To Learn More

ON THE WEB

FACTSURFER

Factsurfer.com gives you a safe, fun way to find more information.

1. Go to www.factsurfer.com.

2. Enter "baby penguins" into the search box and click 🔍.

3. Select your book cover to see a list of related content.

Index

dad, 10, 12
dive, 20
down, 8, 18
eat, 14
eggs, 6
feathers, 18, 19
fish, 14
food, 12
grow, 18

hatch, 6, 7
huddle, 16, 17
hungry, 12
mom, 10, 12, 14
penguin, 4
safe, 10, 16
swim, 20
warm, 8, 16
water, 6

The images in this book are reproduced through the courtesy of: Shutterstock, front cover; Eric Isselee, pp. 3, 22 (life stages--all); Juan Martinez, p. 4; Roger Clark ARPS, pp. 5, 10-11; webguzs, p. 6; Tui De Roy/ SuperStock, pp. 6-7; vladsilver, p. 8; Samantha Crimmin, pp. 8-9, 16-17, 22 (huddle); sabine_lj, p. 12; Tasfoto/ Alamy, pp. 12-13; Graham, David/ Alamy, pp. 14-15; Bambara, p. 18; Sergey 402, pp. 18-19, 22 (hatch); Foto 4440, p. 20; Colin Monteath/ SuperStock, pp. 20-21; Philipp Salveter, p. 22 (eat); Guy Cowdry, p. 22 (swim); fieldwork, p. 23 (down); ValerieVSBN, p. 23 (feathers); JeremyRichards, p. 23 (huddle).